50

MATH
REASONING
QUESTIONS

VOLUME 1

E. A. MORA

Copyright © 2022 E. A. Mora

All rights reserved.

If you find typographical errors, format, syntax errors, logic errors, accuracy, and/or other errors of any kind, please contact the author.

Corrections and improvements are welcome.

This version was prepared by E. A. Mora.

Website: https://linktr.ee/eamora.

Email: eamora.freelancer@gmail.com.

DEDICATION

To my parents.

Thanks for all your hard work and for always being there for me.

MATH REASONING PROBLEMS VOL. 1

CONTENTS

DEDICATION	iii
CONTENTS	v
PREFACE	ix
QUESTIONS	11
QUESTION 1	11
QUESTION 2	11
QUESTION 3	12
QUESTION 4	12
QUESTION 5	13
QUESTION 6	13
QUESTION 7	14
QUESTION 8	14
QUESTION 9	15
QUESTION 10	15
QUESTION 11	16
QUESTION 12	16
QUESTION 13	17
QUESTION 14	17
QUESTION 15	17
QUESTION 16	18
QUESTION 17	18
QUESTION 18	19
QUESTION 19	19
QUESTION 20	20
QUESTION 21	20
QUESTION 22	21
QUESTION 23	21
QUESTION 24	22
QUESTION 25	22
QUESTION 26	23

QUESTION 27	23
QUESTION 28	24
QUESTION 29	24
QUESTION 30	25
QUESTION 31	25
QUESTION 32	26
QUESTION 33	26
QUESTION 34	27
QUESTION 35	27
QUESTION 36	28
QUESTION 37	28
QUESTION 38	29
QUESTION 39	29
QUESTION 40	30
QUESTION 41	30
QUESTION 42	31
QUESTION 43	31
QUESTION 44	32
QUESTION 45	32
QUESTION 46	33
QUESTION 47	34
QUESTION 48	34
QUESTION 49	35
QUESTION 50	35
ANSWER KEY	37
SOLUTIONS	39
SOLUTION TO QUESTION 1	39
SOLUTION TO QUESTION 2	39
SOLUTION TO QUESTION 3	40
SOLUTION TO QUESTION 4	40
SOLUTION TO QUESTION 5	41
SOLUTION TO QUESTION 6	41
SOLUTION TO QUESTION 7	42
SOLUTION TO QUESTION 8	42
SOLUTION TO QUESTION 9	42

SOLUTION TO QUESTION 10	43
SOLUTION TO QUESTION 11	44
SOLUTION TO QUESTION 12	44
SOLUTION TO QUESTION 13	45
SOLUTION TO QUESTION 14	45
SOLUTION TO QUESTION 15	46
SOLUTION TO QUESTION 16	47
SOLUTION TO QUESTION 17	47
SOLUTION TO QUESTION 18	47
SOLUTION TO QUESTION 19	48
SOLUTION TO QUESTION 20	49
SOLUTION TO QUESTION 21	50
SOLUTION TO QUESTION 22	51
SOLUTION TO QUESTION 23	51
SOLUTION TO QUESTION 24	52
SOLUTION TO QUESTION 25	52
SOLUTION TO QUESTION 26	53
SOLUTION TO QUESTION 27	53
SOLUTION TO QUESTION 28	54
SOLUTION TO QUESTION 29	54
SOLUTION TO QUESTION 30	54
SOLUTION TO QUESTION 31	55
SOLUTION TO QUESTION 32	55
SOLUTION TO QUESTION 33	56
SOLUTION TO QUESTION 34	56
SOLUTION TO QUESTION 35	57
SOLUTION TO QUESTION 36	57
SOLUTION TO QUESTION 37	57
SOLUTION TO QUESTION 38	58
SOLUTION TO QUESTION 39	58
SOLUTION TO QUESTION 40	58
SOLUTION TO QUESTION 41	59
SOLUTION TO QUESTION 42	59
SOLUTION TO QUESTION 43	60
SOLUTION TO QUESTION 44	60

SOLUTION TO QUESTION 45	61
SOLUTION TO QUESTION 46	61
SOLUTION TO QUESTION 47	61
SOLUTION TO QUESTION 48	62
SOLUTION TO QUESTION 49	62
SOLUTION TO QUESTION 50	62
ABOUT THE AUTHOR	63

PREFACE

Since high school, I have dedicated special attention to math reasoning problems.

I always found them very appealing, entertaining, challenging, and engaging.

During my studies, I collected more than 5000 of these problems, from all levels.

This short book is a compilation of 50 of those problems.

You only need to know basic arithmetic and algebra and have a pencil and paper to solve these problems.

The problems are not ordered according to their difficulty.

I got the problems presented in this book from a professor or a classmate.

These problems were rewritten many times during the different courses where they were applied.

Names, values, and many other things are different from the original problems.

If for any reason, you or someone you know claims that any of the problems presented here are related to them by copyrights, just send me a message.

MATH REASONING PROBLEMS VOL. 1

QUESTIONS

QUESTION 1

Monica has 27 capybaras in a shelter.

This year, all but 13 capybaras have had babies.

None of them has had more than two baby capybaras.

23 baby capybaras have been born in total.

How many of the baby capybaras have had one baby only?

A. 3
B. 5
C. 6
D. 1

QUESTION 2

A factory has 3 assembly lines.

The number of items in line 2 equals the sum of items in line 1 and line 3.

At a certain time of the day, 5 items are sent from line 3 to line 2 and 7 items are sent from line 2 to line 1.

Also, 6 items from line 1 leave the factory.

How many more items are now in line 2 than in line 1 and line 3 combined?

A. 6 more items
B. 3 more items
C. 2 more items
D. 1 more item

QUESTION 3

Four Olympic athletes, Nadia, Mary, Jessica, and Naomi, have surnames Allen, Armstrong, Barnes, and Campbell - but not necessarily in that order.

i. Jessica can swim faster than Allen but can't run as fast as Barnes.
ii. Allen is a faster runner than Naomi but can't swim as fast as Armstrong.
iii. Nadia is faster than both Jessica and Campbell but can't run as fast as Allen.

Based on that information, what is the surname of each athlete?

A. Nadia is Armstrong, Mary is Campbell, Jessica is Barnes, and Naomi is Allen.
B. Nadia is Barnes, Mary is Allen, Jessica is Armstrong, and Naomi is Campbell.
C. Nadia is Campbell, Mary is Allen, Jessica is Armstrong, and Naomi is Barnes.
D. Nadia is Allen, Mary is Barnes, Jessica is Campbell, and Naomi is Armstrong.

QUESTION 4

In a basketball tournament, the blue team and the red team are disputing the championship.

The blue team scored one-sixth of its points in the 1st quarter, one-third of its points in the 2nd quarter, one-fourth of its points in the 3rd quarter, and the remaining points in the 4th quarter.

The blue team won by 2 points.

The combined score of the two teams playing was 190 points.

How many points did the blue team score in the 4th quarter?

A. 24 points
B. 22 points
C. 23 points
D. 21 points

QUESTION 5

An online computer store is selling hard drives of certain colors at discount.

An online order from three customers is received and Robert, the head of the packaging department, starts to process it.

At a certain point, Robert realized he put the wrong address labels on each of the parcels.

Josef ordered two graphite hard drives, Nick ordered two silver hard drives, and Teresa ordered one graphite and one silver hard drive.

Each pair of hard drives has been packed separately in identical boxes before being wrapped up into the customers' parcels.

How many pairs of hard drives will Robert have to inspect before he can correctly relabel the parcels?

A. 3
B. 0
C. 1
D. 2

QUESTION 6

A dancing academy offers two monthly plans depending on how immersive will be the training.

Plan 1, the most intense preparation program, costs $200 per month. Plan 2, costs $150 per month.

The number of dancers that chose plan 2 increased from 30 dancers to the same number of dancers that chooses plan 1.

If the institute received $11850 for one day, what is the total number of dancers that paid for the institute plans?

A. 68
B. 46
C. 72
D. 53

QUESTION 7

On a planet far away from the earth, if an inhabitant is over fifty, he cannot travel to Alpha Centauri.

Everyone of importance on that planet is either traveling to Alpha Centauri or the Cartwheel Galaxy or both.

Only if one is allowed to travel to Needle Galaxy can one travel to Cartwheel Galaxy.

You can't both not be under forty and be allowed to travel to Needle Galaxy.

An inhabitant called Cygnus is someone of importance on that planet and is not under forty.

Cygnus's age is:

A. Under 50 years old
B. Between 40 and 50 years old
C. More than 50 years old
D. Can't be determined

QUESTION 8

Stephanie and Gina live in the same building.

Gina's apartment is sixteen stories above Stephanie's apartment.

One day, Stephanie was going up the stairs to visit Gina, and halfway up the stairs, she was on the fourteenth floor.

What floor does Gina live on?

A. 21st floor
B. 23rd floor
C. 27th floor
D. 22nd floor

QUESTION 9

In a small star far away from the earth, there are three types of inhabitants: Dracos, Craters, and Orions, indistinguishable between them and with different standards of veracity.

Dracos always tell the truth. Craters always lie.

The Orions, when asked a series of questions, tell the truth and lie alternately; their first answer in a series of questions may be either true or otherwise.

One native of the star, named Eridanus, is the subject of a discussion. The following statements are known:

i. Mr. Draco states: "I'm not a Draco. Eridanus is a Crater."
ii. Mr. Crater states: "I'm not Crater. Eridanus is an Orion."
iii. Mr. Orion states: "I'm not Orion. Eridanus is a Draco."

Which type of inhabitant is Eridanus?

A. Draco
B. Crater
C. Orion
D. Aquila

QUESTION 10

Jonas bought 600 avocados at $5 a dozen.

During the transport, 15 avocados got damaged.

After that, Jonas decided that, when selling the avocados, for every dozen, he will give one avocado for free.

How much should he sell each dozen so that the total profit is $110?

A. $7
B. $4
C. $3
D. $8

QUESTION 11

An iguana needs to travel 1000 meters.

To do this, the iguana will first run 200 meters at the rate of 8 meters per second, then will walk 200 meters at the rate of 2 meters per second, and then repeat this process until it has gone 1000 meters.

How many seconds will the iguana need to travel the full distance?

A. 275 seconds
B. 175 seconds
C. 257 seconds
D. 375 seconds

QUESTION 12

In a pipe factory, the cutting machine got damaged due to overheating.

Six employees volunteered to manually cut large plastic pipes into 0.5-meter tubes to meet the demand on time.

The three leaders of the pairs were Alex, Ramon, and Joshua.

Alex and Miguel would cut up 2-meter tubes, Ramon and Cristian, 1.5-meter tubes, and Joshua and Victor, 1-meter tubes. (These are all first names.)

On the next day, the factory praised the good work of the teams led by Smith, Perez, and Brooks.

Smith and Wilkinson cut tubes into 26 small tubes, Perez and Phillips, 27, and Brooks and Reid, 28. (These are all last names.)

What is Phillips's first name?

A. Cristian
B. Miguel
C. Joshua
D. Ramon

QUESTION 13

Monica, Dayana, and Veronica complete a 100-meter race.

Monica finished the race 10 meters ahead of Dayana, and, subsequently, Dayana finished the race 10 meters ahead of Veronica.

Assuming all three ran at constant speeds, how far ahead was Monica over Veronica when she finished the race?

A. 19 meters
B. 14 meters
C. 15 meters
D. 11 meters

QUESTION 14

In a group of capybaras, the two lightest capybaras weigh 25% of the total weight of the group.

The three heaviest capybaras weigh 60% of the total weight of the group.

How many capybaras are in the group?

A. Three capybaras
B. Five capybaras
C. Six capybaras
D. Four capybaras

QUESTION 15

Given two different numbers, we know that three times the greater exceeds twice the less by 294, and the sum of twice times the greater and five times the less is 481.

What are those numbers?

A. 111 and 22
B. 101 and 31
C. 115 and 12
D. 128 and 45

QUESTION 16

In a 2-hour Spanish test, there are 220 questions, including 44 questions of advanced grammar.

Twice as much time is given to each advanced grammar question as to the other questions.

How much time is allotted to the 44 advanced grammar questions?

A. 36 minutes
B. 41 minutes
C. 35 minutes
D. 48 minutes

QUESTION 17

A Judo dojo is waiting for a famous athlete to arrive.

A vehicle from the dojo was sent by the sensei to meet the athlete at the airport.

The plane landed ahead of schedule, and the athlete was taken toward the dojo in a taxi.

After half an hour the taxi met the dojo vehicle on the road and the athlete was transferred.

The dojo's vehicle returned to the dojo 20 minutes before it was expected.

How many minutes early did the plane land?

A. 45 minutes
B. 40 minutes
C. 25 minutes
D. 30 minutes

QUESTION 18

Jerome, Malcolm, and Lou, a judge, a lawyer, and a prosecutor, not necessarily in that order, and no one has two professions at the same time, are having a meeting to discuss a case, all they sat at a rounded table.

Each person gives a folder to the person sitting at his right hand.

Malcolm gave a green folder to the judge.

Jerome gives a yellow folder to the person that gave a folder to the lawyer.

What is the name of the prosecutor?

A. Lou
B. Jerome
C. Malcolm
D. Ralph

QUESTION 19

Kate was hired as the chef apprentice in a famous restaurant.

Since she is still learning, his boss, to make some pressure on her, decided the following: He will pay her $20 for each dish prepared correctly and fine her $10 for each dish wrongly prepared.

At the end of 30 dishes, neither owes anything to the other.

How many dishes did Kate prepare correctly?

A. 10 dishes
B. 11 dishes
C. 21 dishes
D. 15 dishes

QUESTION 20

A dog owner has a total of 70 dogs, between Corgis, Pomeranians, and Bulldogs.

If he had 10 more Bulldogs, 3 more Pomeranians, and 5 fewer Corgis, he would have an equal number of dogs of each breed.

How many Corgis does the farmer have?

- A. 21 Corgis
- B. 36 Corgis
- C. 45 Corgis
- D. 31 Corgis

QUESTION 21

100 ants are searching for flowers to eat.

43 ants prefer Bee Balm flowers, 35 prefer Cardinal Flowers, and 42 prefer Zinnia flowers.

12 ants prefer Cardinal and Zinnia, 15 prefer Bee Balm and Cardinal, 17 prefer Bee Balm and Zinnia, and 7 prefer all three leaf types.

How many ants prefer none of these leaves?

- A. 11 ants
- B. 17 ants
- C. 15 ants
- D. 10 ants

QUESTION 22

A teacher is going to apply a test to his students.

The questions in the test are of the type: true or false.

The test has only five questions and the students know that:

i. There are more false questions than true ones.
ii. There are not three questions in a row with the same truth value
iii. The first and the last have opposite values.

How many questions are true and how many are false?

A. 3 False and 2 True
B. 2 True and 2 False
C. 1 False and 1 True
D. 2 True and 3 False

QUESTION 23

Luka is selling his pistachio ice cream on the internet.

Since he is starting his business and wants to attract clients, he has a promotion for six people, two ice creams for each one.

One day, he noticed that the amount of blend he prepared will make only one and one-half ice cream for each person.

For each of the persons to have exactly two ice creams, he must add a certain amount of blend to the original amount.

How much blend must Luka add to the original amount?

A. 1/7
B. 1/4
C. 1/3
D. 1/5

QUESTION 24

David, in his new home, has a closet with 6 empty compartments.

David also has 3 baskets filled with the same number of blankets.

He puts 12 blankets from each basket in each compartment.

Each basket has now 12 more blankets than each compartment.

How many blankets did each basket have at the beginning?

A. 48 blankets
B. 18 blankets
C. 98 blankets
D. 68 blankets

QUESTION 25

A biology teacher proposes an experiment to his students.

They are to keep a spider in a special glass jar for exactly 9 minutes.

The teacher only gives the students 4-minute and 7-minute hourglasses with which to measure time, no other time-measuring instruments are allowed.

To complete the experiment, the spider can be in the jar at small intervals of time, but the total time that it should be in the glass jar must be 9 minutes.

The student or students that complete the experiment will have extra credit in the biology class.

What is the shortest time the experiment can be completed?

A. 12 minutes
B. 8 minutes
C. 2 minutes
D. 9 minutes

QUESTION 26

On the New York/Chicago train are three passengers named Spencer, Michael, and Aiden.

By coincidence, the minister, the ophthalmologist, and the announcer have the same last names.

i. Passenger Spencer lives in New York.
ii. The announcer lives halfway between New York and Chicago.
iii. The passenger with the same last name as the announcer lives in Chicago.
iv. The passenger who lives nearest to the announcer earns exactly two times as much a month as the announcer.
v. Passenger Michael earns $3115 a month.
vi. Aiden (a member of the crew) recently beat the ophthalmologist at Racquetball.

What is the minister's last name?

A. Aiden
B. Spencer
C. Michael
D. Joel

QUESTION 27

In a Frisbee tournament, 48 teams are participating.

4 points are disputed in each game, and everyone plays against everyone.

When there is a tie, the four points in dispute are distributed (2 points for each team).

What was the maximum number of tied matches, if the absolute champion got 98 points?

A. 1126
B. 1080
C. 1127
D. 2048

QUESTION 28

A private animal exhibit has come to Brisbane.

In there, seventeen owners of black-footed ferrets are present.

Some of the owners have 1 black-footed ferret, and some have 2 black-footed ferrets.

Counting the legs of the black-footed ferrets and the legs of the owners, there is a total of 126 legs.

All black-footed ferrets have 4 legs and all owners have 2 legs.

How many black-footed ferrets' owners have two black-footed ferrets?

A. 5
B. 3
C. 6
D. 2

QUESTION 29

Norman bought a lot of shirts for $1800 for every 100 shirts.

He sold them for $240 a dozen, earning $6000 in the business.

How many hundreds did the lot have?

   A. 320
   B. 300
   C. 280
   D. 180

QUESTION 30

Dana has $32 to buy cookies for her dogs.

If she buys $5 cookies, she would lack money.

If she buys the $4 cookies, she would have money left over.

How many dogs does Dana have?

A. 5 dogs
B. 7 dogs
C. 6 dogs
D. 4 dogs

QUESTION 31

Matthew sold a bass guitar to Sandy for 20% off the purchase price.

One year later, Sandy sold the bass guitar to Jake for 25% off the price she paid to Matthew.

If Matthew purchased the bass guitar for $750, what is the total percentage discount from Matthew to Jake?

A. 40%
B. 15%
C. 25%
D. 10%

QUESTION 32

Barrel A and barrel B weigh 44 kilos together.

Barrel B and barrel C weigh 48 kilos together.

All 3 barrels weigh 75 kilos together.

What is barrel A's weight?

A. 12 kg
B. 24 kg
C. 18 kg
D. 27 kg

QUESTION 33

An ant and a cockroach decide to race.

The cockroach advances 2 meters every second.

The ant advances 1.5 meters every second.

The cockroach gives the ant a 15-meter head start in a race and reaches the finish line exactly 3 seconds before the ant.

How many meters did the cockroach run?

A. 33 meters
B. 60 meters
C. 98 meters
D. 78 meters

QUESTION 34

To celebrate the retirement of some colleagues, twelve teachers went out to a club.

After the celebration ended, the account totaled $828.

Each one of the non-retired teachers paid an additional $23 to cover the expenses of the retired ones.

How many of them were retired?

A. 3
B. 5
C. 4
D. 6

QUESTION 35

A standard twelve-hour clock shows a time of $10:45$ now.

What time will the clock show 100 hours from now?

A. $2:35$
B. $9:44$
C. $2:45$
D. $1:14$

QUESTION 36

Marcia and Lorena are Soccer pins collectors.

If Lorena gave Marcia 10 of his Soccer pins, they would have the same number of Soccer pins.

However, if Marcia were to give Lorena 10 of his Soccer pins instead, then Lorena would have 5 times as many Soccer pins as Marcia would have.

What is the total number of Soccer pins both Marcia and Lorena started with?

A. Marcia: 20 Soccer pins. Lorena: 40 Soccer pins
B. Marcia: 25 Soccer pins. Lorena: 45 Soccer pins
C. Marcia: 20 Soccer pins. Lorena: 20 Soccer pins
D. Marcia: 40 Soccer pins. Lorena: 20 Soccer pins

QUESTION 37

At a clothing store, two jerseys and two jogging pants cost $85.

A set of two Polo shirts and two Jogging pants cost $120.

A set of a jersey, a polo shirt, and jogging pants cost $75.

What is the cost of the jogging pants?

A. $27.5
B. $17.5
C. $45.5
D. $32.5

QUESTION 38

Victoria went to her music lesson yesterday.

After she finish the lesson, she bought a latte and a sesame cookie.

The price of the latte is $3 plus half its price.

The price of the sesame cookie is $1.5 plus half its price.

She bought one latte and one sesame cookie.

What was the total cost if the tax is included?

A. $9
B. $7
C. $4
D. $6

QUESTION 39

Lady Pink, Lady Purple, and Lady Cyan had lunch together.

One wore a pink headband, one a purple headband, and the other, a cyan headband.

"It's funny", said the one with the purple headband.

"The color of the headbands we wear corresponds to our surnames, but none of us wear the color of our surname."

"In fact, you are right", replied Lady Pink.

What color was Lady Cyan's headband?

A. Purple
B. Cyan
C. Orange
D. Pink

QUESTION 40

A quail egg seller covers his clients' demands with 120 eggs per day he gets from his 8 quails.

Now, the egg demand has increased to a point where he will need 40 more eggs per day.

How many quails of the same production will he have to add to the ones he already has to cover the new demand?

A. 11 quails
B. 10 quails
C. 12 quails
D. 19 quails

QUESTION 41

Two hummingbirds depart from two points A and B, distant 200 km between them. They go to meet them.

The hummingbird coming from A has a speed of 40 km/h.

The hummingbird coming from B, which parted 2 hours earlier, has a speed of 30 Km/h.

In how much time after A's hummingbird departure will they meet?

A. 1 hour
B. 2 hours
C. 4 hours
D. 5 hours

QUESTION 42

In a math training center, the instructors carried out a survey, asking the students if they would rather practice with the American Math Olympiad questions or with the International Math Olympiad questions.

68 students said they'd like to practice with the American Math Olympiad questions.

33 students said either of both was fine.

12 students said they'd prefer International Math Olympiad questions.

If 127 students were asked, how many didn't give their opinion?

A. 57
B. 67
C. 17
D. 47

QUESTION 43

Hallie is enjoying her vacations in Tahiti.

She noticed that it got cloudy on thirteen days.

When it got cloudy in the morning the afternoon was fine.

Every cloudy afternoon was preceded by a fine morning.

There were eleven fine mornings and twelve fine afternoons.

How long were her vacations?

A. 18 days
B. 19 days
C. 11 days
D. 12 days

QUESTION 44

The lifesavers on an Australian beach received a help request from another beach.

They need to go in their ship down the coast, to reach the other beach.

Going with the current they can cover the 2 km in 4 minutes.

Returning against the current, which is steady, will take them 8 minutes.

How long will it take for the lifesavers, in still water, when there is no current?

A. 3 minutes and 15 seconds
B. 2 minutes and 50 seconds
C. 5 minutes and 20 seconds
D. 1 minute and 30 seconds

QUESTION 45

Sixth-grade boys and girls are asked about the drink they prefer between water, soda, and juice.

Of the 68 students asked, 26 prefer the water, and 9 of them are boys.

14 boys prefer juice and 6 of the 37 girls like soda.

How many girls prefer water and how many prefer juice?

A. 14 prefer water and 21 prefer juice
B. 21 prefer water and 8 prefer juice
C. 17 prefer water and 14 prefer juice
D. 20 prefer water and 15 prefer juice

QUESTION 46

There are a group of octopuses with 6, 7, and 8 tentacles in a certain town at the bottom of the sea.

Those with 7 tentacles always lie.

Those with 6 or 8 tentacles always tell the truth.

One day, four octopuses met.

The blue octopus said: "Between the four of us we have a total of 28 tentacles."

The green octopus said: "Between the four of us we have a total of 27 tentacles."

The yellow octopus said: "Between the four of us we have a total of 26 tentacles."

The red octopus said: "Between the four of us we have a total of 25 tentacles."

How many tentacles does the red octopus have?

A. 7
B. 8
C. 6
D. Can't be determined

QUESTION 47

Madeleine was walking past a fashion store and noticed a strange fact:

a) All but three of the dresses on sale were made by Gucci,
b) All but three of the dresses were made by Prada,
c) All but three were made by Burberry,
d) All but three were made by Hermes.

What's the smallest number of dresses that Madeleine could have seen in the store?

A. 5 dresses
B. 2 dresses
C. 4 dresses
D. 8 dresses

QUESTION 48

A captain in the Caribbean was surrounded by a group of sea serpents, from which:

a) 3 could nothing to the right,
b) 3 could see nothing to the left,
c) 3 could see to the right,
d) 3 could see to the left,
e) 3 could see both to the right and the left
f) 3 were blind.

What is the minimum number of snakes necessary for all these circumstances to occur with them?

A. 1 snake
B. 2 snakes
C. 3 snakes
D. 4 snakes

QUESTION 49

There are six moles' nests in the garden of a house.

Each nest has less than four moles.

24 moles are living in those nests.

What is the least possible number of moles in any one nest?

A. 8 moles
B. 9 moles
C. 5 moles
D. 6 moles

QUESTION 50

In the USA, a slow bus departed from New York at 8:00 a.m. and arrived in Boston at 5:00 p.m.

An express bus departed from New York later and arrived in Boston at the same time as the slow bus.

The express bus was four times as fast as the slow bus.

Each bus traveled at a constant speed and did not stop along the way.

When did the express bus depart from New York?

A. 2:45 pm
B. 3:00 pm
C. 2:30 pm
D. 5:30 pm

MATH REASONING PROBLEMS VOL. 1

ANSWER KEY

1. B
2. C
3. B
4. A
5. C
6. C
7. B
8. D
9. A
10. D
11. A
12. A
13. A
14. C
15. D
16. D
17. B
18. A
19. A
20. D
21. B
22. A
23. C
24. A
25. D
26. A
27. C
28. C
29. B
30. B
31. A
32. D
33. D
34. A
35. C
36. A
37. A
38. A
39. A
40. B
41. B
42. D
43. A
44. C
45. C
46. A
47. C
48. A
49. B
50. A

SOLUTIONS

SOLUTION TO QUESTION 1

Since 13 of the capybaras did not have babies, the remaining $27 - 13 = 14$ capybaras had at least one baby.

If they had had exactly one baby each, there would have been 14 baby capybaras in total.

Thus, the remaining $23 - 14 = 9$ babies must have come from capybaras having two babies each, since none of the capybaras had more than two babies.

Therefore, 9 capybaras had two babies, and $14 - 9 = 5$ capybaras had one baby.

The correct choice is option B.

SOLUTION TO QUESTION 2

When 5 items go from line 3 to line 2 there will be 10 more items in line 2 than in lines 1 and 3 combined (line 3 sends 5 items, line 2 receives 5 items).

When 7 items go from line 2 to line 1, there will be 4 more items in lines 1 and 3 combined than in line 2.

After the 6 items get out of the factory, there will be 2 more items in line 2 than in line 1 and line 3 combined.

The correct choice is option C.

SOLUTION TO QUESTION 3

From (iii), we know Nadia can't be Campbell or Allen.

We also know from (i) that Jessica can't be Allen or Barnes, and from (iii) that he cannot be Campbell.

That means he must be Armstrong and Nadia must be Barnes.

From (ii), Allen is a faster runner than Naomi, which means Mary must be Allen, leaving Naomi to be Campbell.

The correct choice is option B.

SOLUTION TO QUESTION 4

With a combined score of 190 points and the blue team winning by 2 points, the score of the game was

$$\frac{190}{2} = 95 + 1 = 96;$$
$$95 - 1 = 94;$$

96 to 94.

In the first quarter, the blue team scored 1/6 of 96 which is 16 points.

In the second quarter 1/3 of 96 which is 32 points.

In the third quarter 1/4 of 96 which is 24 points.

Therefore, the blue team scored the remaining

$$96 - (16 + 32 + 24) = 24 \text{ points}$$

in the fourth quarter.

The right choice is option A.

SOLUTION TO QUESTION 5

Robert must look for the parcel wrongly addressed to Teresa and open it.

If he sees a graphite pair of hard drives then he knows that this parcel must be the one for Josef.

It follows that, as every parcel was wrongly addressed, the one addressed to Josef should go to Nick, and the one addressed to Nick should go to Teresa.

The same analysis follows if the pair of hard drives were found to be silver.

Hence, Robert will only need to have a look at one pair of hard drives to be able to decide how to readdress the packages.

The correct choice is option C.

SOLUTION TO QUESTION 6

For the 30 more students that choose plan 2, the institute received: $150 \times 30 = \$4500$.

For an equal number from each plan, the institute would have received: $\$11850 - \$4500 = \$7350$.

For 1 student in each plan, the institute receives:

$$\$200 + \$150 = \$350.$$

Students that chose plan 1:

$$\frac{\$7350}{\$350} = 21.$$

Students that chose plan 2:

$$21 + 30 = 51.$$

The total number of students paying for plans 1 and 2 is

$$21 + 51 = 72.$$

The correct choice is option C.

SOLUTION TO QUESTION 7

Cygnus is not under 40, then he is not allowed to travel to Needle Galaxy.

He cannot travel to Cartwheel Galaxy.

Since he is someone of importance, he can travel to Alpha Centauri.

He cannot be over 50.

Hence, Cygnus is between 40 and 50 years old.

The correct choice is option B.

SOLUTION TO QUESTION 8

At the halfway point, Stephanie was on the fourteenth floor.

She had $22 - 14 = 8$ floors to go.

Then, Gina lives on the $14 + 8 = 22^{nd}$ floor.

The correct choice is option D.

SOLUTION TO QUESTION 9

Only one Orion can say that he's not Draco.

For a Draco, if he said so, would be lying; and a Crater would be telling the truth.

Thus, Mr. Draco is the Orion; and Mr. Orion, who says he is not an Orion, must be the Draco.

Hence, he tells the truth about Eridanus.

Eridanus is a Draco.

The correct choice is option A.

SOLUTION TO QUESTION 10

The number of dozens purchased:
$$\frac{600}{12} = 50.$$

Amount of investment:
$$50 \times \$5 = \$250.$$

The goal is to recover the investment and earn $110.

For this, Jonas must collect:
$$\$250 + \$110 = \$360.$$

The avocados that are suitable for sale are:
$$600 - 15 = 585.$$

Since 1 avocado is given for free for every dozen sold, the avocados must be sold in 13 by 13.

Number of dozens sold = $\frac{585}{13} = 45.$

For earning $360 in the sale of 45 dozen, each dozen must be sold at:
$$\frac{\$360}{45} = \$8.$$

The correct choice is option D.

SOLUTION TO QUESTION 11

The iguana needs $\frac{200}{8} = 25$ seconds to run the first 200 meters and $\frac{200}{2} = 100$ seconds to walk the next 200 meters.

For the first 400 meters, it will require 125 seconds. (1000 − 400 = 600 meters remain)

For the next 400 meters, it will require another 125 seconds. (600 − 400 = 200 meters remain)

For the final 200 meters, while running, it will require 25 seconds to cover them.

The iguana will require a total of $125 + 125 + 25 = 275$ seconds to travel the 1000 meters.

The correct choice is option A.

SOLUTION TO QUESTION 12

Cutting 2-meter tubes produces a number of 0.5-meter tubes divisible by 4.

Cutting 1.5-meter tubes, divisible by 3.

1-meter tubes, divisible by 2.

Phillips's team's 27 tubes are not divisible by 2 or 4, so it is the 1.5-meter-tube team of Ramon and Cristian.

The team leader is Ramon Perez, so Phillips's first name is Cristian.

The correct choice is option A.

SOLUTION TO QUESTION 13

In the time it takes Dayana to complete the final 10 meters, Veronica would have run less than 10 meters, so, at the instant Monica finished, Veronica must have been closer than $10 + 10$ meters.

When Monica completes the 100-meter race, Dayana is 10 meters behind, so he has run 90 meters in the same amount of time. This means Dayana runs at 90 percent the speed of Monica ($D = 0.9M$).

Similarly, when Dayana completes the 100-meter race, Veronica is 10 meters behind, so he has run 90 meters at the same time. Thus, Veronica runs at 90 percent of the speed of Dayana ($V = 0.9D$).

Combining those equations, we find that Veronica runs at 81 percent the speed of Monica ($V = 0.9D = 0.9[0.9M] = 0.81M$).

Then, when Monica runs 100 meters to complete the race, in that time Veronica would have run 81 meters.

Therefore, Veronica is 19 meters behind Monica when she finishes.

The correct choice is option A.

SOLUTION TO QUESTION 14

Taking away the two lightest and the three heaviest, the remaining capybaras weigh 15% of the total.

But if there were two or more, then two of them would weigh at most 15% of the total, which is less than the two lightest ones, which is not possible.

Hence, there is one more capybara and there is a total of six capybaras.

The correct choice is option C.

SOLUTION TO QUESTION 15

Step 1. Formulate the equations.

Let x be the greater number and y be the less number.
Then

$$\text{EQ1: } 3x - 2y = 294$$
$$\text{EQ2: } 2x + 5y = 481$$

Step 2. Get the value of y.

Multiply EQ1 by -2, and EQ2 by 3,

$$-2 \times \text{EQ1} = -6x + 4y = -588$$
$$3 \times \text{EQ2} = 6x + 15y = 1443$$

Add, algebraically, the new EQ1 and the new EQ2 and solve for y

$$6x + 15y = 1443$$
$$-6x + 4y = -588$$
$$\overline{19y = 855}$$
$$y = 45$$

Step 3. Get the value of x.

Substitute the value of y in EQ1 or EQ2 (the one you find more convenient) and solve for x

$$3x - 2(45) = 294$$
$$3x - 90 = 294$$
$$3x = 384$$
$$x = 128$$

The correct choice is option D.

SOLUTION TO QUESTION 16

Count each advanced grammar question as 2 units: $2 \times 44 = 88$ questions.

In terms of time, the advanced grammar questions represent $\frac{88}{220}$ or $\frac{2}{5}$ the total time.

Since 2 hours = 120 minutes, $\frac{2}{5} \times 120 = 48$ minutes.

The correct choice is option D.

SOLUTION TO QUESTION 17

The dojo's vehicle would have taken 20 minutes to go from where it met the taxi to the airport and back.

Thus, it was 10 minutes from the airport when it met the taxi.

These 10 minutes plus the 30 minutes the taxi had been driving before they met made 40 minutes the plane was ahead of schedule.

The correct choice is option B.

SOLUTION TO QUESTION 18

Let's suppose Jerome is on the left side of Malcolm and Lou is on the right side of Malcolm. Then, Malcolm gives a folder to Lou, thus Lou will be the judge.

Since Malcolm is on the right side of Jerome, Malcolm gives a folder to the lawyer. Since Lou can't be the judge and the lawyer at the same time, then our previous assumption is not correct.

Then, Jerome must be on the right side of Malcolm, and Lou on the left side of Malcolm. Thus, Jerome is the judge that gives a folder to Lou and Lou gives a folder to the lawyer that must be Malcolm.

Hence, Lou is the prosecutor.

The correct choice is option A.

SOLUTION TO QUESTION 19

Let x be the number of dishes prepared correctly, and y be the number of dishes prepared wrongly.

Then:
$$x + y = 30$$
$$\text{and } 20x - 10y = 0$$

From the first equation: $y = 30 - x$

Substituting the value of y in the first equation gives us
$$20x - 10(30 - x) = 0$$

Solving for x we have
$$20x - 300 + 10x = 0$$
$$20x + 10x - 300 = 0$$
$$30x = 300$$
$$x = \frac{300}{30} = 10$$

The correct choice is option A.

SOLUTION TO QUESTION 20

Let's denote Corgis as C, Pomeranians as P, and Bulldogs as B.

Then, we have
$$C + P + B = 70,$$
and
$$(C - 5) + (P + 3) + (B + 10) = (70 + (10 + 3 - 5)) = 78.$$

We know that $(C - 5) = (P + 3) = (B + 10)$.

Let's denote that same number as x.

So,
$$x + x + x = 78$$
$$3x = 78$$
$$x = 78/3$$
$$x = 26.$$

Hence, the number of Corgis the farmer has is
$$C - 5 = 26$$
$$C = 26 + 5$$
$$C = 31$$

The correct choice is option D.

SOLUTION TO QUESTION 21

Since 7 ants prefer all three leaf types and 17 prefer Bee Balm and Zinnia, there must be $17 - 7 = 10$ who prefer only Bee Balm and Zinnia, but not Cardinal.

There are 8 that prefer only Bee Balm and Cardinal, and 5 prefer only Cardinal and Zinnia.

43 ants prefer Bee Balm, and 10, 7, and 8 prefer the possible combinations of Bee Balm with at least one other type, so there are $43 - 10 - 7 - 8 = 18$ that prefer only Bee Balm.

There are 20 that prefer only Zinnia and 15 that prefer only Cardinal.

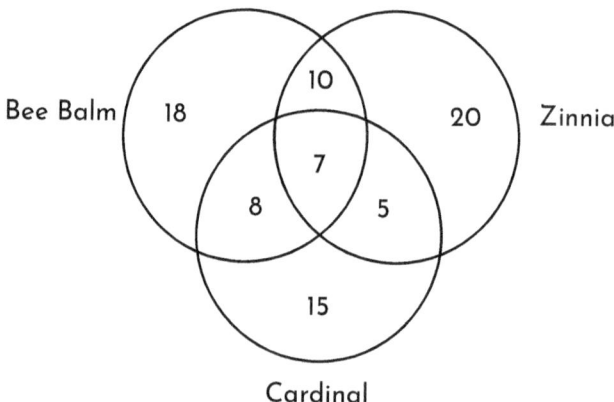

Thus, the number of ants that prefer at least one leaf is the sum of the numbers in all the regions of the diagram, which is 83.

Hence, the number of ants that prefer no leaves is

$$100 - 83 = 17.$$

The correct choice is option B.

SOLUTION TO QUESTION 22

Five questions must be answered with true or false.

The possible answers are:

a. 5 false and 0 true
b. 4 false and 1 true
c. 3 false and 2 true
d. 2 false and 3 true
e. 1 false and 4 true
f. 0 false and 5 true

Possibilities d, e, and f are eliminated by (i).

Possibility a is eliminated by (ii) or (iii).

Only possibilities b and c remain.

The possibility b is eliminated by (ii) and (iii).

The only feasible possibility is c.

It cannot be eliminated, since it agrees with (i), (ii), and (iii).

Hence, there are 3 false and 2 true.

The correct choice is option A.

SOLUTION TO QUESTION 23

The current blend will make $6 \times 1½ = 9$ ice creams.

He must make 3 more ice creams so that each person will have 2 ice creams.

If x is the needed amount to make 3 more ice creams, then:

$$9 \times x = 3$$
$$x = \frac{3}{9} = \frac{1}{3}.$$

He must add 1/3 of the original amount to make 12 ice creams.

The correct choice is option C.

SOLUTION TO QUESTION 24

Each compartment gets $3 \times \frac{1}{12} = ¼$ of a filled basket of blankets.

Each basket loses $6 \times \frac{1}{12} = ½$ of a filled basket.

Each basket has $½ - ¼ = ¼$ of a filled basket more than each compartment.

Therefore, 12 blankets represent ¼ of a filled basket.

A filled basket contained $4 \times 12 = 48$ blankets at the beginning.

The correct choice is option A.

SOLUTION TO QUESTION 25

At time 0, turn over both hourglasses.

At time 4, turn over 4 min hourglass when it finishes.

At time 7, turn over 7 min hourglass when it finishes.

At time 8, as the 4 min hourglass finishes, turn over 7 min hourglass again (it will have measured 1 minute).

At time 9, take out the spider when the 7 min hourglass finishes.

The experiment can be done in 9 minutes.

The correct choice is option D.

SOLUTION TO QUESTION 26

The passenger who lives nearest to the announcer is not Michael (iv-v).

He does not live in New York or Chicago, since at best these are only tied for nearest to the announcer (ii), so he is not Spencer (i).

By elimination, he is Aiden.

Since the passenger from Chicago is not Spencer (i), by elimination he is Michael, and the announcer's name is Michael (iii).

Since Aiden is not the ophthalmologist (vi), by elimination he is the minister.

The correct choice is option A.

SOLUTION TO QUESTION 27

Analyzing the 48 teams and their 47 matches each, the total number of matches is $\frac{48 \times 47}{2} = 1128$.

Then, the champion in his 47 matches scored 98 points.

Since we want to maximize the number of tied matches, these 98 points could be obtained with 47 games tied (94 points) and one won (4 points).

To keep the maximum number of tied matches, we consider that all other parties were tied.

Hence, the largest number of tied matches is 1127.

The correct choice is option C.

SOLUTION TO QUESTION 28

The 17 black-footed ferrets' owners have a total of 34 legs, so the ferrets must have a total of $126 - 34 = 92$ legs.

Since all ferrets have 4 legs, there must be $92/4 = 23$ ferrets and so there must be $23 - 17 = 6$ owners with two ferrets.

The correct choice is option C.

SOLUTION TO QUESTION 29

The selling price per unit was: $240/12 = \$20$.

The selling price per hundred is $20 \times 100 = \$2000$.

Since the cost per hundred was $1800, the profit on a hundred shirts is $\$2000 - \$1800 = \$200$.

If the total profit was $6000, the number of hundreds that Norman sold was: $\$6000/\$200 = 300$.

The right choice is option B.

SOLUTION TO QUESTION 30

If she buys the $5 cookies, she can buy: $\$32/\$5 = 6.4$ cookies.

Since she will lack money with this option, this means that Dana has more than 6 dogs.

If she buys the $4 cookies, she can buy: $\$32/\$4 = 8$ cookies.

Since she will have money left over with this option, this means that Dana has fewer than 8 dogs.

Then, the number of dogs is between 6 and 8, so, Dana has exactly 7 dogs.

The correct choice is option B.

SOLUTION TO QUESTION 31

Matthew paid $750 and sold it to Sandy for 20 percent off of $750

$$\$750 \times 0.20 = \$150$$
$$\$750 - \$150 = \$600$$

Sandy then sold it to Jake for 25 percent off $600

$$\$600 \times 0.25 = \$150$$
$$\$600 - \$150 = \$450$$

The total percentage discount from Matthew to Jake was

$$\$750 - \$450 = \$300$$
$$\frac{300}{750} = \frac{2}{5} = 0.4 = 40\%$$

The correct choice is option A.

SOLUTION TO QUESTION 32

Since all 3 barrels weigh 75 kilos and barrel A and barrel B weigh 44 kilos, barrel C must weigh $75 - 44 = 31$ kilos.

Since barrel B and barrel C weigh 48 kilos together, barrel B must weigh: $48 - 31 = 17$ kilos.

Then

$$A + B + C = 75kg$$
$$A + 17kg + 31kg = 75kg$$
$$A + 48kg = 75kg$$
$$A = 75kg - 48kg$$
$$A = 27kg$$

The correct choice is option D.

SOLUTION TO QUESTION 33

The cockroach gains 0.5 meters every second, so it will take it 30 seconds to overcome the 15-meter lead.

Since it finishes 3 seconds ahead of the ant, it has a 4.5-meter lead at the finish.

It would take it 9 seconds to gain 4.5 meters on the ant.

In all, the cockroach ran for $30 + 9 = 39$ seconds.

$$2 \times 39 = 78 \text{ meters}$$

The correct choice is option D.

SOLUTION TO QUESTION 34

Let be x the number of retired teachers that were invited.

If each one of the teachers on the celebration put money to pay the account, each one would have to put: $828/12 = $69.

The number of teachers that paid is: $12 - x$ and each one put $69 + $23 = $92 and with this amount, it must be totalized the $828.

Then
$$828 = (12 - x) \times \$92$$
$$828 = \$1104 - \$92x$$
$$92x = \$1104 - \$828$$
$$x = \frac{\$276}{\$92}$$
$$x = 3$$

Hence, there were 3 retired/invited teachers.

The correct choice is option A.

SOLUTION TO QUESTION 35

At the end of 96 hours (8 groups of 12 hours each), the clock will still show 10:45, with 4 hours left over from the original 100 hours.

4 hours after 10:45 is 2:45.

The correct choice is option C.

SOLUTION TO QUESTION 36

If, when Lorena gave Marcia 10 Soccer pins, they had an equal amount, Lorena must have started with 20 more Soccer pins than Marcia.

Therefore, when Marcia gave Lorena 10 Soccer pins, she would have 40 more Soccer pins than Marcia.

When looking for two numbers that are 40 apart where the larger is 5 times the smaller, 10 and 50 are the only two numbers that satisfy those statements.

Therefore, Marcia started with 20 Soccer pins and Lorena with 40 Soccer pins.

The correct choice is option A.

SOLUTION TO QUESTION 37

All 3 together cost $75.

One jersey and one jogging pant cost: $85/2 = $42.5.

Therefore, the Polo shirt must cost $75 − $42.5 = $32.5.

Since one Polo shirt and one jogging pant cost $120/2 = $60 and the Polo shirt is $32.5.

Then, the jogging pant must cost $60 − $32.5 = $27.5.

The correct choice is option A.

SOLUTION TO QUESTION 38

If the latte costs $3 plus half its price, then the $3 must be the "other half" of the price.

So, the latte costs $6.

Similarly, the price of the sesame cookie must be $1.5 plus $1.5 or $3.

The total cost is $6 + $3 = $9.

The correct choice is option A.

SOLUTION TO QUESTION 39

Lady Pink could not wear the pink headband since it would correspond to her surname.

Lady Pink couldn't wear the purple one either, because the one with the purple headband was talking to her.

Thus, Lady Pink's headband had to be cyan.

The purple and pink headbands remain, which were worn respectively by Lady Cyan and Lady Purple.

Hence: Lady Pink has the Cyan headband, Lady Purple has the Pink headband, and Lady Cyan has the Purple headband.

The correct choice is option A.

SOLUTION TO QUESTION 40

The target production of each quail is: $120/8 = 15$ *eggs*.

To meet the additional demand are necessary: $40/8 = 5$ more quails.

The correct choice is option B.

SOLUTION TO QUESTION 41

B's hummingbird has already traveled 30 km × 2 = 60 km when A's hummingbird goes out.

Then the distance between them is 200 km − 60 km = 140 km.

Every hour they approach each other 40 km + 30 km = 70 km.

They will meet in 140 km/70 km = 2 hours after the departure of A's hummingbird.

The correct choice is option B.

SOLUTION TO QUESTION 42

68 students chose American Math Olympiad questions.

33 students did think any of the options were fine.

Then, 68 − 33 = 35 students want only American Math Olympiad questions.

Thus, 12 + 33 + 35 = 80 students that gave their opinion.

Hence, 127 − 80 = 47 students did not give their opinion.

The correct choice is option D.

SOLUTION TO QUESTION 43

There are three possible types of the day:

a) Clouds in the morning and fine in the afternoon.
b) Fine in the morning and clouds in the afternoon.
c) Fine in the morning and fine in the afternoon.

Let the number of such days in each category be a, b, and c.

The number of days on which got cloudy: $a + b = 13$.

The number of days having fine mornings: $b + c = 11$.

The number of days having fine afternoons: $a + c = 12$.

From these equations, we derive that $a = 7$, $b = 6$, and $c = 5$.

The number of days on vacation is $7 + 6 + 5 = 18$.

The correct choice is option A.

SOLUTION TO QUESTION 44

Speed downstream is ½ km per minute.

The return speed is ¼ kilometer per minute.

Therefore, the current makes 1/8 of a kilometer difference per minute.

Consequently, the ship's speed is 3/8 of a kilometer per minute, which translates into 5 ⅓ minutes or 5 minutes and 20 seconds, for the 2 kilometers in still water.

The correct choice is option C.

SOLUTION TO QUESTION 45

If 37 of the 68 asked students are girls, then, the remaining 31 are boys.

If of the 31 children, 9 prefer water and 14 juice, then, 8 prefer soda.

If 8 boys and 6 girls prefer soda, a total of 14 prefer this drink.

Girls that prefer water: $26 - 9 = 17$.

Boys and girls that prefer juice: $68 - (26 + 14) = 28$.

Girls that prefer juice: $28 - 14 = 14$

The correct choice is option C.

SOLUTION TO QUESTION 46

If the red octopus told the truth then, to add up to 25, there should be three octopuses with 6 tentacles and one with 7.

That is, two more octopuses, apart from the red one, would have told the truth, which is impossible. Therefore, the red octopus lies and has 7 tentacles.

Moreover, it can be seen that the yellow one has 6 tentacles and the others 7.

The correct choice is option A.

SOLUTION TO QUESTION 47

Let's focus our attention on what need not be there instead of what must be in the store.

There is at least one dress made by each brand, thus there need to be at least four dresses in the store.

We need one by each of the given brands. The other 3 are not by that brand. There need be only four dresses in the store.

The correct choice is option C.

SOLUTION TO QUESTION 48

There were three snakes blind and three with both eyes healthy.

There were at least 6 snakes.

The correct choice is option A.

SOLUTION TO QUESTION 49

To find the least number of moles possible in one nest, it is necessary to find the maximum number of moles in the other 5 nests.

Five nests can contain at most: $5 \times 3 = 15$.

Hence, there are at least $24 - 15 = 9$ moles in the remaining nest.

The correct choice is option B.

SOLUTION TO QUESTION 50

From the departure and arrival times, we can work out that the slow bus took 9 hours to reach Boston.

Since the express bus was four times as fast as the slow bus, it must have taken 2 hours and 15 minutes to reach Boston, and so it departed from New York at 2:45 p.m. (5:00 p.m. minus 2 hours 15 minutes).

The correct choice is option A.

ABOUT THE AUTHOR

I have a bachelor's degree in mathematics.

I completed my university degree in 2014.

Since then, I work as a freelance math content writer, LaTeX typesetter, illustrator, and editorial designer.

www.ingramcontent.com/pod-product-compliance
Lightning Source LLC
Chambersburg PA
CBHW070318220526
45465CB00004B/1904